W9-AXI-614

Jaden Toussaint,
The Greatest

Episode 1

THE QUEST FOR SCREEN TIME

Plum Street Press
A Division of Yes, MAM Creations

Published by **Plum Street Press**
Copyright © 2015 by **Marti Dumas**
All Rights Reserved.

Contents

To My Sunshine—
You make me happy even
when skies are gray.

Prologue

SPEAKS KINDNESS, OOZES CONFIDENCE.

JADEN TOUSSAINT

Specializes in: knowing stuff. And also, ninja dancing. He's really, really good at ninja dancing.

Past president, International Society of Stealthy Felines. Resigned in scandal. Likes to be held like a baby and scratch stuff.

GRANDMASTER, CAT CHESS.

GRIS GRIS

JOKESTER AND ACTION EXPERT.

OWEN

Extreme dinosaur
safari bungee
jumping?
Owen is your guy.

EXCELS AT : BEING IN CHARGE.

EVIE

Don't let the cuteness
fool you. This girl packs a
punch.

THE FRIENDS

CICADA HUNTER AND MATH WHIZ.

SONJA

Also draws
excellent rainbows.

SPOTS HURT FEELINGS AND DISTRACTED GOALIES.

WINSTON

Can quote stats
from every World
Cup final.*
*that he has been alive for

THE FAM

Mama:
Loving. No nonsense. Most often seen reading fantasy books or experimenting with bean desserts. Gives good hugs.

Baba:
Tall. Competitive. Competitive about being tall. Gives great piggy back rides. Prefers to be called "baba" which means "father" in Swahili. Does not speak Swahili.

Sissy:
Reader. Writer. Animal lover. Once gave up meat for 6 months, but was broken by the smell of turkey bacon. Plans to be the 1st PhD chemist to star in a Broadway musical.

x

ANIMAL OF MYSTERY

??????

Guinea Pig never has the same name two weeks in a row.

This week, you may call him Zed.

Chapter 1
THE TROUBLE WITH BOOKS

Jaden Toussaint hated to read.

I mean he really,

totally, honest to goodness

hated reading, and he avoided it whenever possible.

Unless there was a book light involved.
But, who doesn't like book lights?
Right?

The most annoying thing was, everyone in Jaden Toussaint's family was a reader.

His mom read fantasy books.

His sister read about fighting cats and Greek gods.

His father read books about becoming a better basketball, football, and soccer coach despite the fact that he actually programmed software for a living.

The three man defense...

... they could think of no way of getting rid of Smaug...

Why was that annoying?
Because it made Jaden Toussaint feel
lonely.

That was the real trouble with books. All
that attention focused on the pages (and
not on him) made Jaden Toussaint feel
left out.

Grumble. Grumble. Discontent.

Um... guys? Guys?
Is anybody gonna play
with me? Hello?

You speak
from your heart,
young Firepaw.

5

Jaden Toussaint was a people person.

He liked to play
    Toranpu cards

and conduct
experiments

and be
an animal
adventurer.

Everybody knows that you can't do any of that by yourself. Well...you can, but it wouldn't be nearly as much fun.

It was especially not fun when grown-ups kept telling you to tone your creature adventuring down to a dull roar.

Who ever heard of a lion with a dull roar? Ridiculous.

Once, Jaden Toussaint's father had been trying to finish reading *Rugby for Dummies*. Even though they were in the park with lots of other things to do, JT felt *so* lonely.

His father had taken pity on him and let him play with his phone—for 10 whole minutes—while he read the last few pages.

Those 10 minutes were grand. They were glorious. They were better than ice cream. Jaden Toussaint had never felt more brilliant. And for him, that was really saying something.

In the short but oh-so-sweet time the phone was in JT's possession, he added purple chickens to a farm, spelled 17 words, and learned that the name for a group of flamingoes is "flamboyance." A flamboyance of flamingoes. How cool is that?

The sum total of human knowledge was in the palm of his hand! With this smartphone he could do ANYTHING.

Chapter 2
NEED. MORE. SCREEN TIME.

Computer? Smart phone? Smart tv? Tablet? It didn't matter. Jaden Toussaint had tasted his destiny, and he wanted more. Luckily, as a scientist, he had the tools he needed to figure it out. Asking for things over and over had sometimes gotten him what he wanted in the past, so maybe repetition would work. He decided to start with that.

The Repetition Test: Day 1

He asked his sister in the kitchen. He asked his mother in the car. He asked his father in the bedroom. No luck.

Jaden Toussaint, the note-taker, noted these failures in his notebook, but otherwise took no note of them. He pressed on.

The Repetition Test: Day 2

This time he asked his mother in the kitchen, asked his father in the car, and his sister in her bedroom. That didn't work either.

But not to worry! Trying again and again is a part of the scientific method. Every good scientist knows that you might have to try hundreds, even thousands of times before your experiment is a success. And Jaden Toussaint was nothing if not a good scientist.

He experimented with location, tone of voice, and time of day. No matter which variable he tested for, nothing seemed to work.

The Repetition Test: Day 23

Even scientists get discouraged sometimes, and Jaden Toussaint was beginning to think that he might never get screen time again.

But all that changed when Miss Bates stepped in.

	Mama	Papa	Diary
Kitchen			
Car	X	X	X
Living Room	X	X	X
Bathroom	X	X	X
Bedroom	X	X	X
Park	X	X	X
Restaurant	X	X	X
Walking Home	X	X	
Walking to School	X	X	X
Dinner	X	X	X
Breakfast	X	X	X
Kitchen (begging)	X	X	X
Car (begging)	X	X	X
Living Room (begging)	X	X	X
Bathroom (begging)	X	X	X
Bedroom (begging)	X	X	X
Park (begging)	X	X	X
Restaurant (begging)	X	X	X
Walking Home (begging)	X	X	X
Walking to School (begging)	X	X	X
Dinner (begging)	X	X	X
Breakfast (begging)	X	X	X
Kitchen (pleading)	X	X	X
Car (pleading)	X	X	X
Living Room (pleading)	X	X	X
Bathroom (pleading)	X	X	X
Bedroom (pleading)	X	X	X
Park (pleading)	X	X	X
Restaurant (pleading)	X	X	X
Walking Home (pleading)	X	X	X
Walking to School (pleading)	X	X	X
Dinner (pleading)	X	X	X
Breakfast (pleading)	X	X	X

Chapter 3
MISS BATES' CLASS

Miss Bates was the world's most perfect kindergarten teacher.

She played math games.

She did science experiments.

She got all of Jaden Toussaint's jokes.

Once she had
even let them
paint on a
wall. Directly
on the wall.
No paper!

MOST IMPORTANT

A PAIR OF
SLIPPERS!
GET IT?
SLIPPERS!

And at circle time, Miss Bates always let him
sit next to his two best friends—Winston
and Owen. (Evie and Sonja were cool, too,
but they sat on the other side of the rug.)

Just when JT thought kindergarten couldn't
get any better, something truly amazing
happened:

KINDERGARTEN HOMEWORK

For the first half of the school year, kindergarteners didn't get to do homework. That was so unfair! Why let them go to school, then deprive them of homework?

When he had asked his mother why kindergarteners were treated so unfairly, she said, "Well, baby, a lot of the kids have never been to school before. Too many new things at once can feel overwhelming, so the teachers hold back on homework so everyone can get used to school first."

"I know," JT blurted out almost before his mother was done talking. He didn't mean to be rude, but it just popped out like that sometimes. His mother gave him an incredulous look.

"I know!" He repeated insistently.

Pause!
The thing is, he didn't really know. I mean, he knew it by the time he said it, but he hadn't known it before his mother said it. You know? Un-pause.

What Jaden Toussaint really meant to say was, "Thanks, Mom! I didn't know that before, but it makes sense, so I know it now." Actually, he did say all of that. But he said it in his mind. "I know," was just the only part that came out.

"Jaden Toussaint," his mother had sighed, "You know a lot, but you don't know everything."

TRYING TO KEEP IT ALL INSIDE

It irked him when his mother said that. He did know everything. Well, almost everything and that was close enough to count.

Luckily, Miss Bates was so awesome that when she said,

"Guess what everyone? This is the week that we start our kindergarten homework assignments,"

and Jaden Toussaint had gleefully shouted, "I know!" instead of saying that annoying thing his mother always said about him not knowing everything,

Miss Bates said, "I'm sure you do."

Then she smiled such a beautiful smile
that JT was sure that she meant it.

The best part was that the kindergarten
homework sheet was drawn like a tic
tac toe board with assignments in each
square.

Computer time was right in the center.

Chapter 4
THE THRILL OF VICTORY

Miss Bates had given them computer time for homework. Let me repeat that: Computer time. For homework. This was possibly the best thing that had ever happened. Computer time was a kind of screen time, so basically Miss Bates had just given JT a free pass for screen time.

You see, Jaden Toussaint had a big sister, so he already knew all about how homework worked. He pretty much considered himself a homework expert.

Here's how it goes:
1. Kids *have* to do their homework.
2. Parents *have* to let kids do their homework.
3. Therefore, wielding the power of a homework assignment, kids are basically in charge of their parents!!!

His quest had come to a glorious end. This was going to be S-WEET!

Jaden Toussaint checked his backpack to make sure that his kindergarten homework sheet was inside before he lined up to leave. It was there. The center square with computer time seemed to shine brightly, as if Miss Bates had written it with gold. Maybe she had. It was all he could do to not shout out: "Yes! YEEESSSS!" in his best mad scientist voice.

But he was afraid that, even though she liked science, Miss Bates might be frightened of mad scientists.

They were mad after all. So Jaden Toussaint kept his cool and said, "Yes!" in an excited whisper instead.

He checked his backpack twice more on the walk to the kindergarten pick-up spot, and as soon as he saw his father he went barreling forward clutching the homework sheet in his hand.

MAD SCIENTIST

Actually, I prefer misunderstood genius on the cutting edge of science.

"Baba! Baba! Guess what!" JT shouted.

"What, bruh?" his father asked.

"Mbtsedwehftadomputimefoamwrk!"

Sometimes when Jaden Toussaint's super big brain kicked into high gear, he talked so fast that people couldn't understand him.

"Hunh, bruh?" his father said.

"I said, Miss Bates said we have to do computer time for homework!"

That was how he remembered it, anyway. But when he showed his father the homework sheet, it somehow said something different.

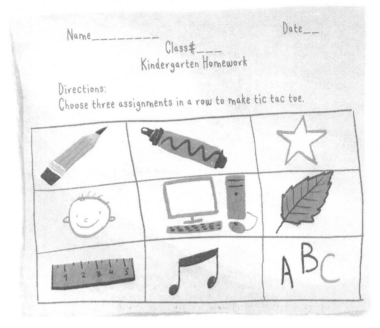

Name_____   Date___
Class#____
Kindergarten Homework

Directions:
Choose three assignments in a row to make tic tac toe.

"Bruh, you know that you don't have to do computer time to do your homework, right?"

"I know," he said.

Pause!

He didn't really know. What he meant to say was, "I didn't know that before. Thank you for pointing it out for me," but he was so disappointed that it came out all wrong.

Wait.
Fry it literally.
Or figuratively.

Unpause.

24

"But Baba," he began, suddenly regaining his confidence. "There are 4 different ways to make tic tac toe with computer time. That's the most! That means Miss Bates thinks it's the most important!"

"Bruh, there are also 4 different ways to make tic tac toe without computer time."

"Nice move, Baba," JT thought. His father was right, but Jaden had learned an important lesson from playing Toranpu so much. Making one good move is not enough to win a whole game. Just like in Toranpu, there is almost always a counter move.

Jaden Toussaint's super-fast brain kicked into gear. Unfortunately, he only had a chance to say, "But, Baba..." before his baba cut him off.

Brain →

"Bruh, your brain is still developing. Did you know that too much screen time will fry your brain?"

"I know," he said.

Pause!

He didn't really know. What he meant to say was, "Oh, no! I didn't know that!" but he shortened it. Sort of.

Unpause.

His father looked incredulous.

"Really! I know, Baba. But since I'm already really smart, doing computer time won't hurt me."

His father seemed to consider it, but in the end he just said, "Pick a way to make tic tac toe with no computer time. But, we can play Toranpu when you're done," before giving Jaden Toussaint an extra tight hug.

Ordinarily the promise of a game of Toranpu would have cured just about anything, but not today. Jaden Toussaint just had to figure out a way to get some more screen time. He just had to.

Chapter 5
PROBLEM SOLVED

Jaden Toussaint's quest for screen time had hit a major snag. The power of homework had failed him. At this rate, it looked like he wouldn't have screen time again until he was a droopy old man.

But did Jaden Toussaint give up? No way!

Having a super big brain had its
advantages.

JT knew that all this problem needed
was some serious thinking. Some super-
powered brain power, if you know what I
mean.

Brain Protection Zone

Screen Time
Blueberries
French
Toranpu
Creature Adventuring
Scientific methods
Sweet Dance Moves
Chinese
Cheese Pizza
Useless Yet Impressive Info
English
In Charge Center
Animal Facts
Swashbuckling
Dr. Hoooo?

Thanks to his gigantic brain, most ideas just popped right into JT's head without him trying. But this screen time thing was a toughie. So far it had evaded all his usual, most scientific methods. There had to be a way, but the solution wasn't popping into his head.

For stubborn problems like this one, there was one surefire way to kick his brain into top gear:

Turn the Page!

33

Then, out of nowhere, it started. That swirly, whirly, zinging feeling he got whenever he was on the verge of a brilliant idea. And just like that, Jaden Toussaint knew what to do.

Chapter 6
THE PETITION

The next morning JT went to school armed with everything he needed: seven sheets of paper taped together and lots of sharpened pencils. The piece of paper was a petition. The pencils were for signing the petition. He had started to bring pens because he had heard somewhere that the pen is mightier than the sword, but they weren't allowed to use pens yet in Miss Bates' class so pencils would have to do.

What we want: A lot more squares of computer time on the homework sheet.

When we want it: As soon as you can make a new homework sheet.

Why we want it: So kids can learn lots of important things from the internet (and also play games) no matter which way they make tic tac toe.

He gathered signatures on the playground before school started and continued at recess. Lots of the kindergarteners were still beginner readers and couldn't read his writing at the top. Not a problem. He read it to them.

Some of the other kindergarteners could read his writing just fine, but had trouble writing their names.

Not a problem. His father had told him stories about the olden times when people like Dred Scott made their mark with an X instead of signing their names.

Jaden Toussaint told the kindergarteners they could sign with any kind of mark they wanted, as long as they would remember it later.

signatures

JT
Eve
Owen

X

Oscar
Kae
Winston

Vincent
Kannon
Zah

Raanan
Ami
Marie

By the end of the day, Jaden Toussaint had gathered 102 signatures-- one from each of the 100 kindergarteners, one from Mr. Alvin, the custodian, and one from Miss Brown who worked in the office.

He presented the petition to Miss Bates while everyone was packing up. Miss Bates was pretty impressed.

"Wow, JT! Did you do all this yourself?" Jaden Toussaint nodded proudly.

"And look at all those signatures! You really worked hard. You should be proud of yourself."

He did feel proud of himself, but he didn't want Miss Bates to think he was too braggy, so he just smiled in response. Owen and Evie gave him a thumbs up.

"Y'all must really love computer time. You especially, Jaden Toussaint," said Miss Bates.

Then Miss Bates smiled her wonderful smile and said, "I can't promise anything, but I would hate to see all your hard work go to waste. Let me see what I can do."

S-WEET!

Chapter 7
A Happy Ending?

When Jaden Toussaint met his mother at the kindergarten pick-up spot, he somehow expected that the world would have magically changed. He imagined his mother greeting him by pulling out her smartphone and asking him if he wanted to play on it while they walked home.

She didn't.

When they made it home and sat down to have a snack, he imagined that his sister would whip out the laptop and type in the passcode. Then the two of them would computer-it-up while munching on popcorn.

That didn't happen either.

A terrible thought suddenly occurred to him. What if Miss Bates didn't do anything?

He didn't think Miss Bates would lie, but maybe she just said she would do something so that he wouldn't be disappointed, even though she already knew that the people at school who were super in charge—even more in charge than the teachers—would never let her do something as huge as change the homework sheet.

That's when he heard a chime. It was a short sound, but JT's keen ears knew it had come from his mother's purse. It was a message chiming on her phone. It was a message from Miss Bates. Nobody told him that's what it was; he just kind of knew.

He watched his mother out of the corner of his eye as she read the message. He tried to be casual about it, but before long he had muscled his way in to try to read along with her. Sometimes mamas don't like it when you try to read their email without permission. Lucky for him, his mama didn't seem to mind... this time.

EMAIL

new message (1)

from:

Miss Bates

Dear JT's Mom,

I just wanted you to know that your son is truly amazing.

Today at school he started a petition for more computer time and got 102 signatures, including 2 adults! That is pretty impressive. The other kindergarten teachers and I have decided to add a bonus choice square to the homework sheet. The bonus square would be for students who feel up to doing an extra assignment. For the bonus, kids will be able to choose any of the nine homework activities to do again, including computer time.

We are so proud of Jaden Toussaint for looking for a way he could solve his problem. Bravo!

Best Always,

Miss Bates
Teacher, the Busy Bees

Do you know what? That did the trick. Jaden Toussaint's mother looked at him with a mixture of love and pride before giving him the biggest bear hug and reminding him that he is her sunshine.

That night, she and his baba let him change his homework tic tac toe to one that included the computer square.

The screen time was awesome, but the hugs were even better.

Surrounded by love this big, he really knew that he was The Greatest.

I love you,
baby.

I love you,
bruh.

I love you,
toots.

I know.
No, Seriously.
I really already
knew that, guys.

Epilogue

Observations- Best Times to Get Extra Screen Time

① What? Smartphone
How Long? 3-15 minutes.
Where? Restaurant
Success rating: 4 out of 10.
After you have entertained and enlightened everyone at the table, but before the food comes, there is a narrow window of time when the grown-ups want to talk about grown up stuff. Use this weakness to your advantage.

② What? Television
How Long? 30-60 minutes.
Where? In the living room on Saturday morning.
Success rating: 7 out of 10
After mom has made coffee, but before it kicks in.

③ What? Computer
How Long? 5-15 minutes
Where? Mom's computer Desk or at the library
Success rating: 10 out of 10
Only if your teacher makes it a part of your homework that you have to do. Or if you hypnotize or use your super powers to get billions of people to sign a petition.
Seriously. Billions.

About the Illustrator

Marie Muravski was born in a city deep in Siberia. There were no bears on the streets of her city, but it was surrounded by the most beautiful pine forests. Those pine forests inspired her to draw as a little girl and she never stopped. Now that she is all grown up, being surrounded by nature still inspires her art.

She and her husband travel the world looking for new beautiful places to inspire them, making art, and just being awesome together.

You can find her at:

www.facebook.com/MarieMuravski

THIS BOOK?·······························

About the Author

Marti Dumas is a mama who spends most of her time doing mama things. You know - feeding ducks in parks, constructing Halloween costumes, facilitating heated negotiations, reading aloud, throwing raw vegetables on a plate and calling it dinner, and shouting, "Watch out!" whenever there are dog piles on the walk to school.

Sometimes she writes, but only very occasionally and in the early morning. And, yes. She really does really, really like fantasy books. A lot.

You can find her at:

www.MartiDumasBooks.com

···

COMING SOON

JADEN TOUSSAINT THE GREATEST

EPISODE 2: DR. HOOOO?

WRITTEN BY MARTI DUMAS
ILLUSTRATED BY MARIE MURAVSKI

AVAILABLE NOW

Ever Dream of **Being a Wolf?**

JALA
and the
WOLVES

by Marti Dumas

For crafts, recipes, and more, visit:

www.MartiDumasBooks.com

Authors love reviews.
We eat them up like pizza for breakfast.

Yum!

CPSIA information can be obtained
at www.ICGtesting.com
Printed in the USA
LVOW12*0125091117

555585LV00008B/94/P